Plants
Important Producers

by Kate Boehm Jerome

Table of Contents

DEVELOP LANGUAGE

This is a garden.
People visit this garden to look at the plants.

Talk about the photos.

How would you describe one of the plants?

The plant has _______.

How are the plants in the photos different?

The plants have different _______.

What do plants need to live?

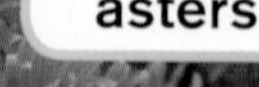

weeping willow trees
tulip
New York Botanical Garden

CHAPTER 1

How Are Plants Grouped?

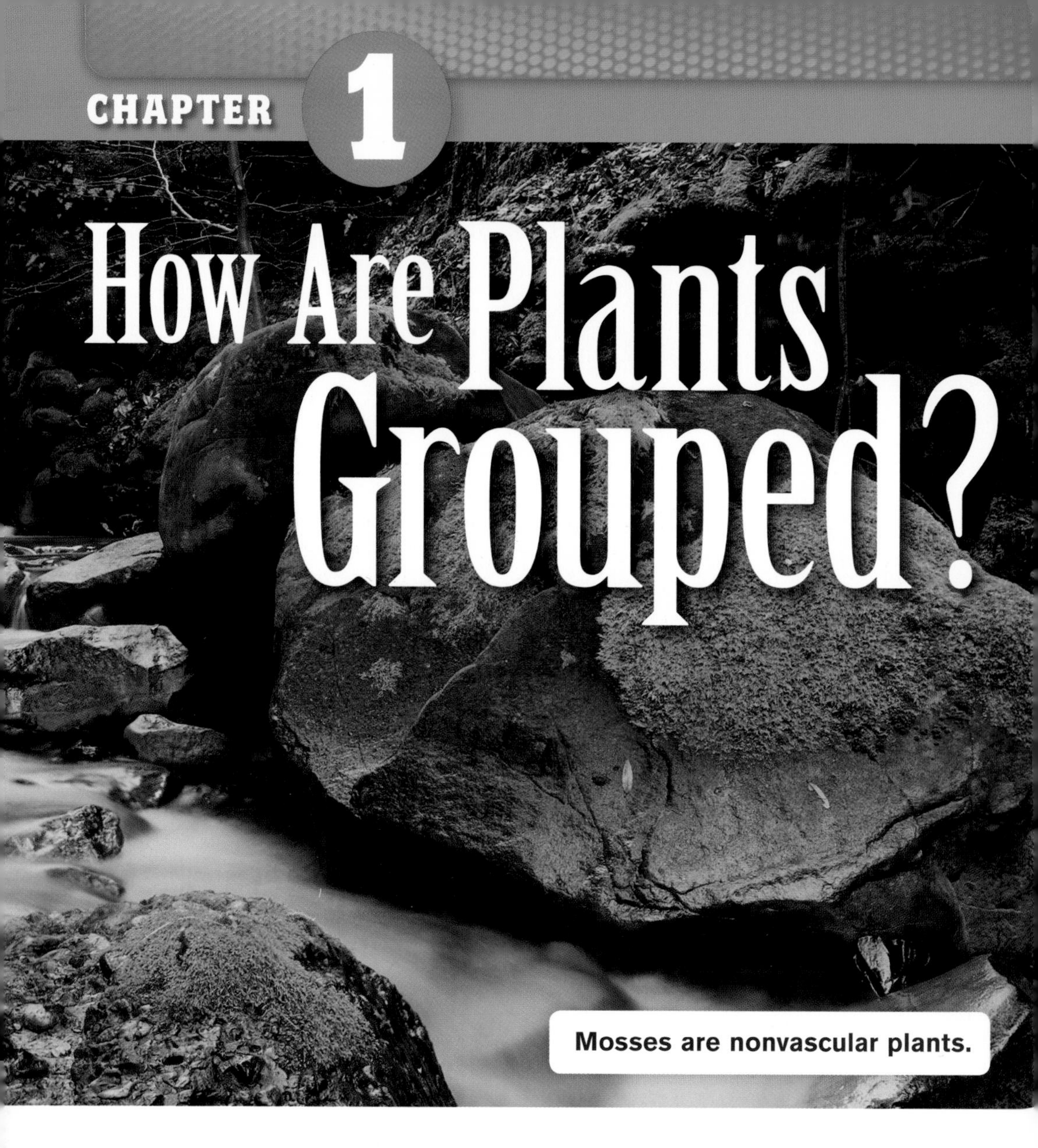

Mosses are nonvascular plants.

There are many kinds of plants.
Each plant belongs to one of two groups.

Nonvascular plants are one group.
These plants cannot grow very tall.
They do not have special **tissues** that move water, minerals, and food through the plant.

Vascular plants are another group.
These plants have special tissues that move water, minerals, and food through the plant.
Vascular plants can grow taller than nonvascular plants.

Redwood trees are tall vascular plants.

KEY IDEA Plants are vascular or nonvascular.

Some kinds of vascular plants produce **seeds**. Each seed can grow into a new plant.

Gymnosperms are vascular plants that produce seeds. They do not produce flowers. The seeds of most gymnosperms form on cones.

Douglas fir trees are gymnosperms.

KEY IDEA Gymnosperms produce seeds but do not produce flowers.

Angiosperms are also vascular plants that produce seeds. Angiosperms produce flowers. Their seeds form inside the flowers.

An apple tree is an angiosperm.

Fruit grows around angiosperm seeds. The fruit protects the seeds.

KEY IDEA Angiosperms produce flowers. Their seeds are protected by fruit.

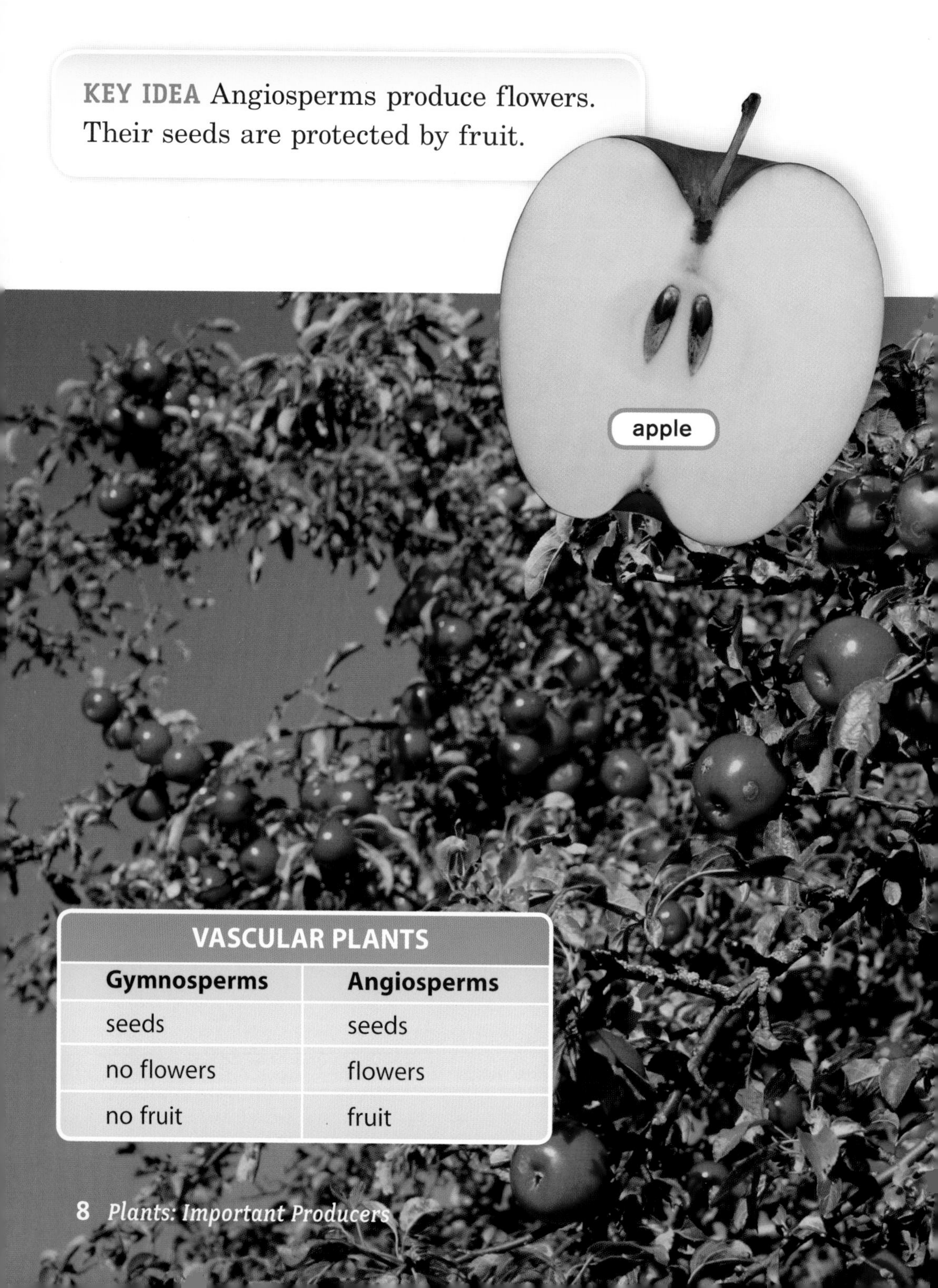

VASCULAR PLANTS	
Gymnosperms	**Angiosperms**
seeds	seeds
no flowers	flowers
no fruit	fruit

YOUR TURN

CLASSIFY

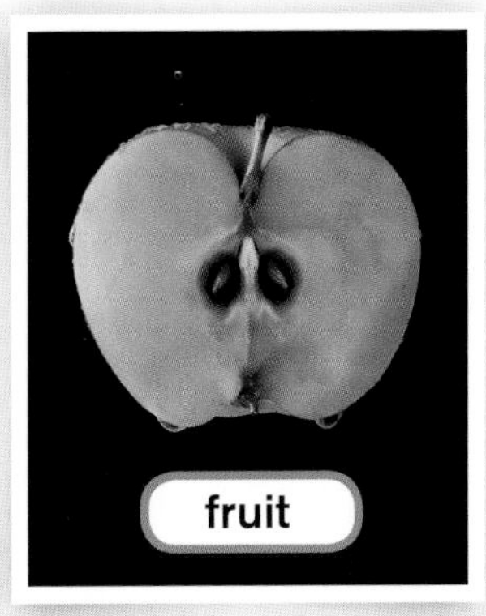

Look at the photos above. Classify each plant part.

Tell if it belongs to a gymnosperm or an angiosperm.

Plant Part	Angiosperm or Gymnosperm
cones	
fruit	
flowers	

MAKE CONNECTIONS

Have you seen trees with cones? What makes those trees different from flowering trees?

USE THE LANGUAGE OF SCIENCE

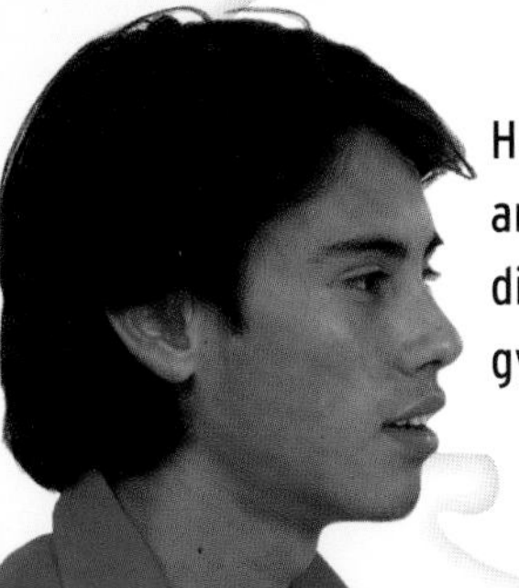

How are angiosperms different from gymnosperms?

Angiosperms produce flowers. Their seeds are protected in fruit.

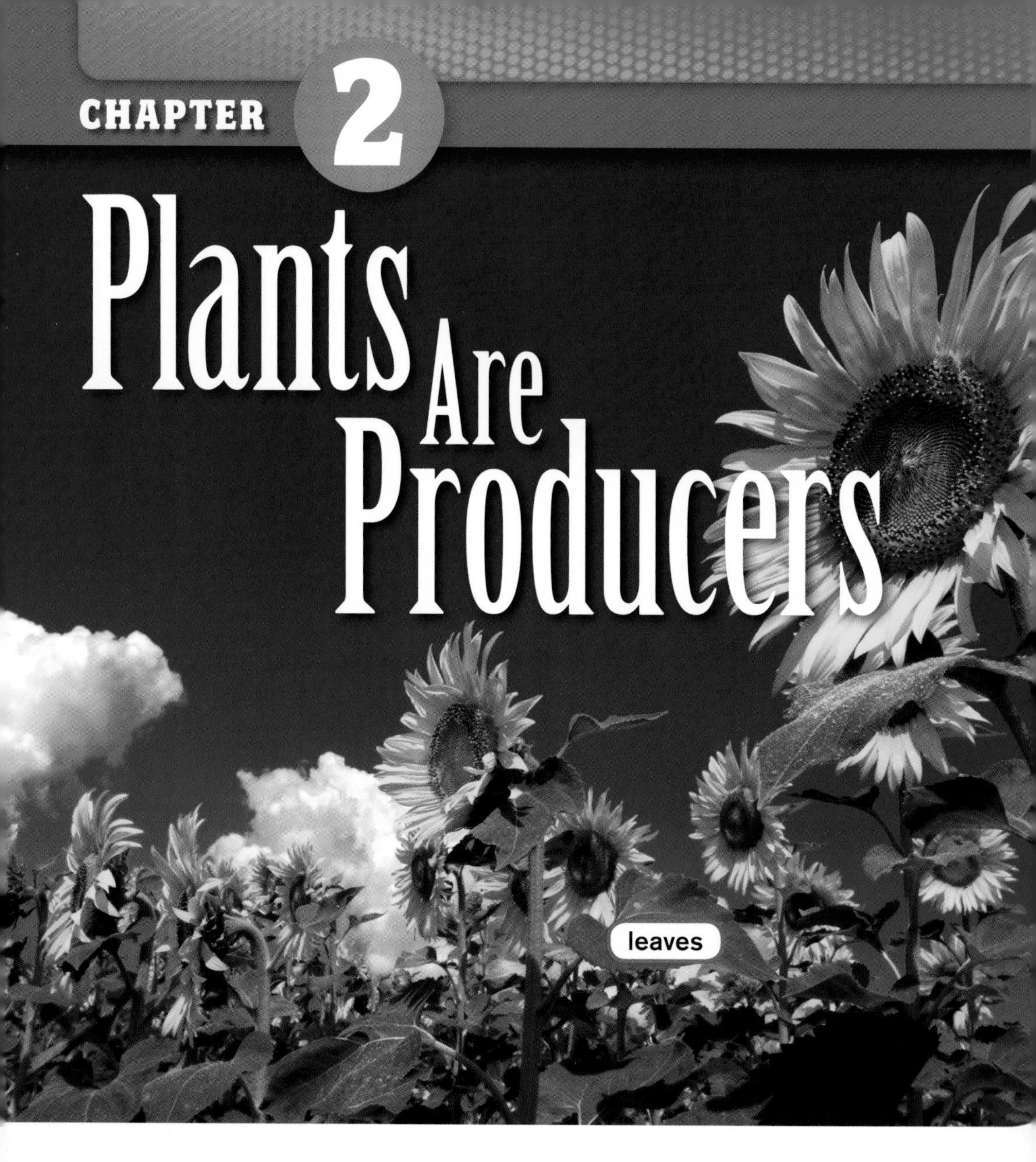

Most plants are **producers**.
This means they make, or produce, their own food.
Plants use energy from sunlight to produce food.

KEY IDEA Plants are producers because they make their own food.

Most plants have roots, stems, and leaves. These help the plants live and grow.

Some plants produce food in their stems. Most plants produce food in their leaves.

▲ A rose makes food in its leaves.

A cactus makes food in its stems.

Plants get energy from the food they make.
Unlike plants, an animal's body cannot make its own food.
Animals must eat other living things to get energy.

Animals that Eat Plants

Some birds eat berries.

Rabbits eat grass.

Cows eat grass.

Animals depend on plants for energy.
Some animals eat plants.
Some animals eat other animals that eat plants.
Without plants, animals would not have any food.

Animals that Eat Other Animals

Wolves eat rabbits.

Some owls eat mice.

Energy moves from plants to animals in a food chain. Plants are important because they are the producers in a food chain.

Explore Language

one wolf, two wolves
one leaf, two leaves

KEY IDEA In a food chain, plants are important because they are producers.

A Forest Food Chain

YOUR TURN

OBSERVE

You know that plants need sunlight to make food. Look at this plant by a sunny window. What can you observe about the plant's leaves?

Most of the leaves are growing toward the ______.

Why do you think the leaves are growing in this direction?

The leaves are growing in this direction because ______.

MAKE CONNECTIONS

Have you seen animals eat plants? Draw and label a picture of an animal eating a plant.

STRATEGY FOCUS

Make Connections

Look back at this chapter. What connections can you make to the words and pictures? Make a chart like this one. Write down your connections.

The text says . . . or The picture shows . . .	This reminds me of . . .
(p. 11) Most plants have roots, stems, and leaves.	*my aunt's garden*

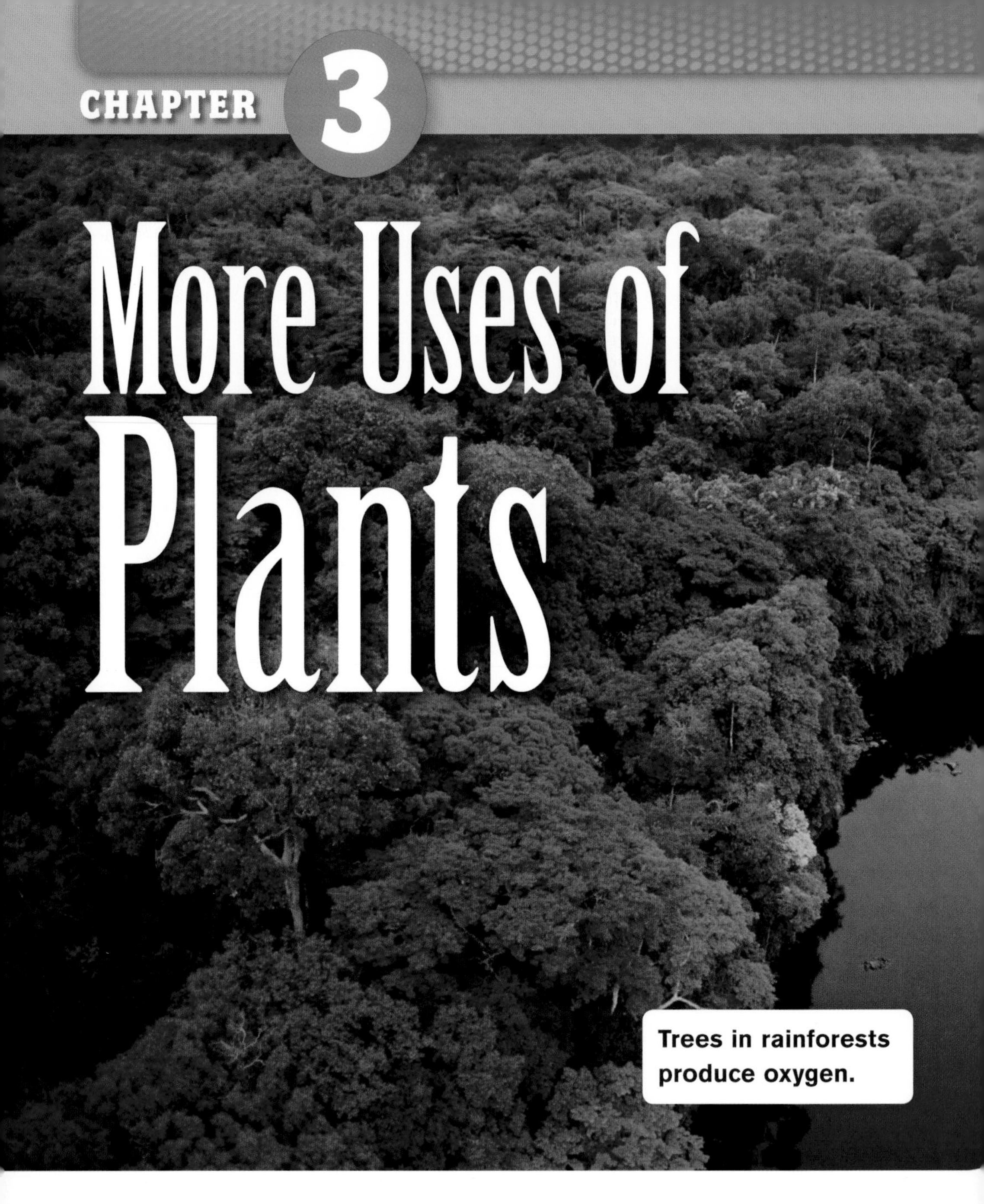

More Uses of Plants

Trees in rainforests produce oxygen.

Plants are important for many reasons.
Plants produce **oxygen** when they make their own food.
The oxygen goes into the air we breathe.
Most living things need this oxygen to live.

People use plants to make many things.
People make paper and clothes from plants.
People use wood from plants to build houses.

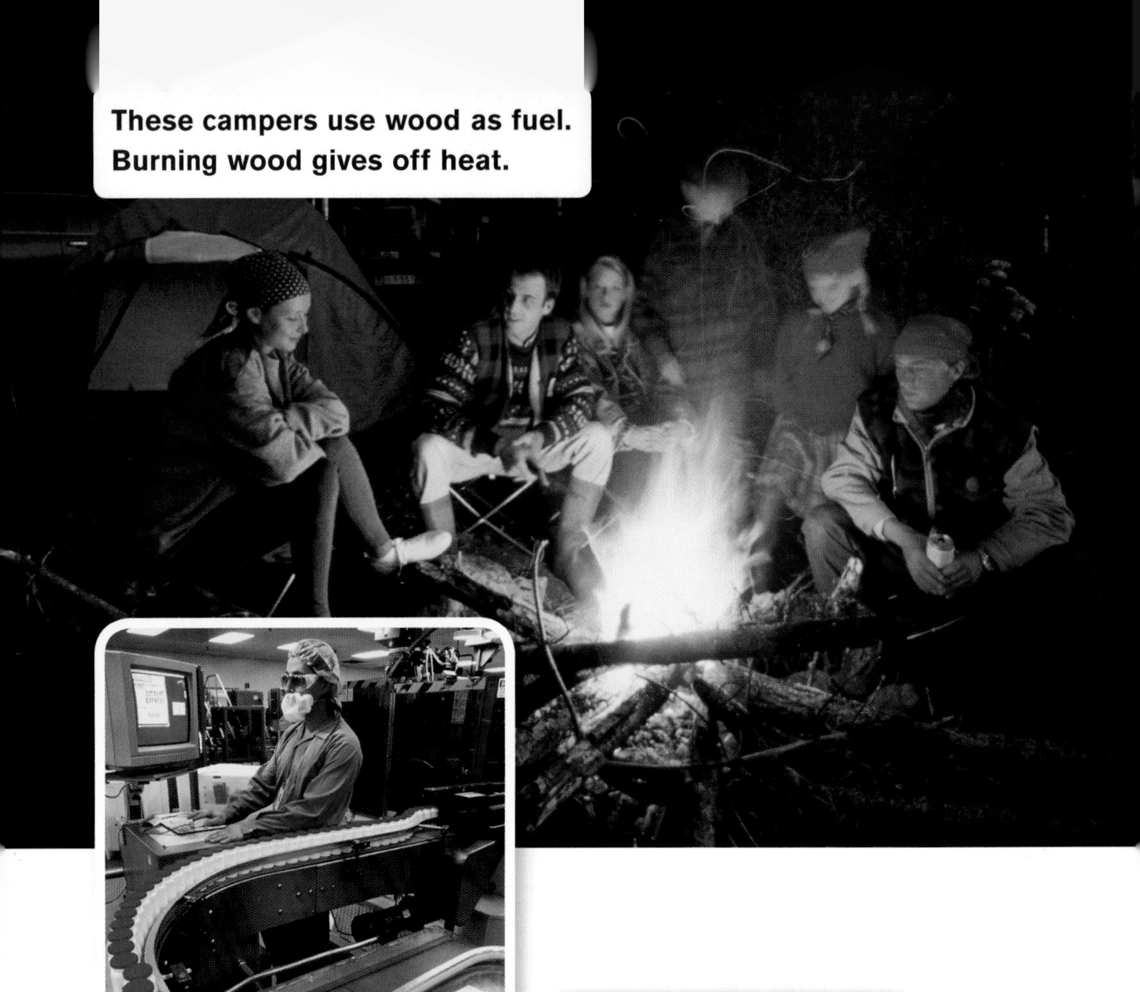

These campers use wood as fuel. Burning wood gives off heat.

Some medicines are made from plants.

People use plants for **fuel**.
People make medicines from plants.
Plants are a very important part of our world!

KEY IDEAS Plants are important for many reasons. Plants produce oxygen and are used to make many things.

YOUR TURN

INFER

Look at the camping photo on page 18.
Tell what you can infer from the photo.

Why do the people want a fire?

The people want a fire because ______.

How are the people using plants?

The people are using plants to ______.

MAKE CONNECTIONS

What things do you use that are made from plants?

EXPAND VOCABULARY

The word **chain** can be used in different ways. Find out what each of these phrases means. Write about it and draw a picture.

supermarket chain **chain-link fence** **chain saw**

How are these things alike? How are they different?

Work with the Land

Do you like working outdoors? Find out about these careers.

◀ natural resource managers

▲ nursery worker

▲ landscape designer

If you like...	then find out about:
nature	natural resource managers
plants	nursery workers
parks	landscape designers

Natural resource managers protect public lands.

Nursery workers grow plants.

Landscape designers draw outdoor plans.

Use Pronouns: *it, its, they, their*

When you describe, you tell what something is like.
You can use pronouns to replace some of the nouns.

When you tell about plants or animals, you can use the subject pronouns **it** and **they**. You can use the possessive pronouns **its** and **their**.

Subject Pronouns: *it, they*	Possessive Pronouns: *its, their*
The rose uses sunlight to make food. **It** uses sunlight to make food.	The rose lost **the rose's** leaves. The rose lost **its** leaves.
Leaves help the rose. **They** help the rose.	Roses have thorns on **the roses'** stems. Roses have thorns on **their** stems.

With a friend, talk about the animals on pages 12 and 13.
Share what you know about these animals. Use **it**, **its**, **they**, and **their**.

Write a Description

Choose a plant that you like. Write sentences to describe it. Tell what the plant is like. Draw and label a picture of your plant. Use nouns and pronouns.

Words You Can Use				
Nouns		**Pronouns**	**Verbs**	**Adjectives**
gymnosperm	stem	it	has, have	vascular
angiosperm	seed	its	grows, grow	nonvascular
cone	flower	they	uses, use	tall
leaves	root	their	produces, produce	beautiful

Plants Help Run Cars

People don't just eat corn. They are using corn and other crops to make ethanol. Ethanol is a clear fuel. It can be mixed with gasoline. Cars that run on this mix of fuels make less pollution.

Read the newspaper article.

- What is ethanol?

Ethanol is ______.

Read the title again. How do plants help run cars?

Plants help run cars by providing ______.

Key Words

angiosperm (angiosperms) a vascular plant that produces seeds and flowers
An apple tree is an **angiosperm**.

gymnosperm (gymnosperms) a vascular plant that produces seeds but no flowers
A pine tree is a **gymnosperm**.

nonvascular plant (nonvascular plants) a plant that does not have special tissues for moving water, minerals, and food through the plant
Moss is a **nonvascular plant** that grows on rocks.

producer (producers) a living thing that uses energy from sunlight to make its own food
Grass is a **producer** that uses sunlight to make food.

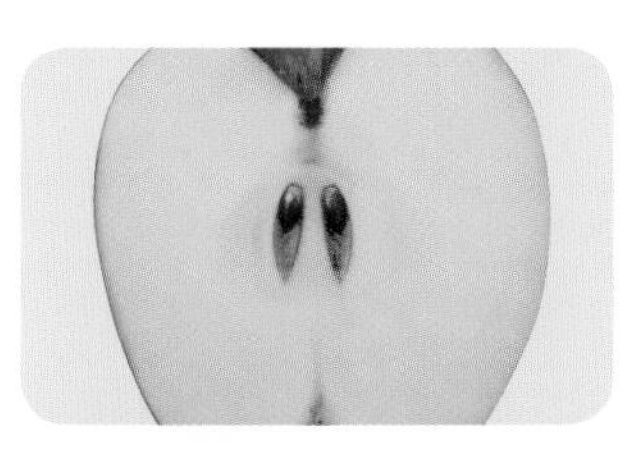

seed (seeds) a part of a plant that can grow into a new plant
Apple **seeds** are protected inside fruit.

vascular plant (vascular plants) a plant with special tissues that move water, minerals, and food through the plant
A redwood tree is a tall **vascular plant**.

Index

MILLMARK EDUCATION CORPORATION
Ericka Markman, President and CEO; Karen Peratt, VP, Editorial Director; Rachel L. Moir, Director, Operations and Production; Mary Ann Mortellaro, Science Editor; Amy Sarver, Series Editor; Betsy Carpenter, Editor; Guadalupe Lopez, Writer; Kris Hanneman and Pictures Unlimited, Photo Research

PROGRAM AUTHORS
Mary Hawley; Program Author, Instructional Design
Kate Boehm Jerome; Program Author, Science

BOOK DESIGN Steve Curtis Design

CONTENT REVIEWER
Nikki L. Hanegan, PhD, Brigham Young University, Provo, UT

PROGRAM ADVISORS
Scott K. Baker, EdD, Pacific Institutes for Research, Eugene, OR
Carla C. Johnson, EdD, University of Toledo, Toledo, OH
Donna Ogle, EdD, National-Louis University, Chicago, IL
Betty Ansin Smallwood, PhD, Center for Applied Linguistics, Washington, DC
Gail Thompson, PhD, Claremont Graduate University, Claremont, CA
Emma Violand-Sánchez, EdD, Arlington Public Schools, Arlington, VA (retired)

PHOTO CREDITS Cover © AGStockUSA, Inc./Alamy; 1 © Holt Studios International Ltd/Alamy; 2-3 © Paul H. Reinert/Alamy; 2a © Michael Shake/Shutterstock; 2b © altrendo nature/Getty Images; 3a © Alexei Noviko/Shutterstock; 3b © cloki/Shutterstock; 4 © Don Smith/Photodisc/Getty Images; 5 © Nathan Jaskowiak/Shutterstock; 6a © Jeff Foott/Discovery Channel/Getty Images; 6b © SCPhotos/Alamy; 7a and 9c © Gordana Sermek/Shutterstock; 7b and 23a © vera bogaerts/Shutterstock; 8a and 23e © Billy Lobo H./Shutterstock; 8b © Michaela Steininger/Shutterstock; 9a © Paul Bodea/Shutterstock; 9b © Phil King/Alamy; 9d and 9e Lloyd Wolf for Millmark Education; 10 © János Gehring/Shutterstock; 11a © ene/Shutterstock; 11b © Carolina K. Smith, M.D./Shutterstock; 11c © Mike Norton/Shutterstock; 12-13 © Alan Majchrowicz/Photographer's Choice/Getty Images; 12a © Andreas Nilsson/Shutterstock; 12b © John Swithinbank/Alamy; 12c © Stijn Peeters/Shutterstock; 13a © Dynamic Graphics Group/Creatas/Alamy; 13b © Mike Rogal/Shutterstock; 14a © as-foto/Shutterstock; 14b © Vladimir Ivanov/Shutterstock; 14c © Joel Sartore/National Geographic/Getty Images; 14d © Tom Brakefield/Digital Vision/Getty Images; 15 © John Kaprielian/Photo Researchers, Inc.; 16 © Edward Parker/Alamy; 17a © Norman Chan/Shutterstock; 17b © Soundsnaps/Shutterstock; 17c © Ryan Klos/Shutterstock; 18a © David De Lossy/Photodisc Green; 18b © Mitch Kezar/Stone/Getty Images; 19 © ajt/Shutterstock; 20a © Edwin Verin/Shutterstock; 20b © Comstock Royalty Free Photograph; 20c © Peter Jordan/Alamy; 21 © Carolina K. Smith, M.D./Shutterstock; 22 © Konstantin Sutyagin/Shutterstock; 22 © Konstantin Sutyagin/Shutterstock; 23b © Peter Blazek/Shutterstock; 23c © Robyn Mackenzie/Shutterstock; 23d © Dmitry Kosterev/Shutterstock; 23f © Chee-Onn Leong/Shutterstock; 24 © Michal Kram/Shutterstock

Published by Millmark Education Corporation
7272 Wisconsin Avenue, Suite 300
Bethesda, MD 20814

ISBN-13: 978-1-4334-0042-1
ISBN-10: 1-4334-0042-1

Printed in the USA

10 9 8 7 6 5 4 3 2 1